Charles S. Philips

Tobacco Curing and Resweating for Quality and Dark Colors

A practical hand-book for cigar manufacturers and leaf dealers who are

licensed to use the patents of Charles S. Philips

Charles S. Philips

Tobacco Curing and Resweating for Quality and Dark Colors
A practical hand-book for cigar manufacturers and leaf dealers who are licensed to use the patents of Charles S. Philips

ISBN/EAN: 9783337042691

Printed in Europe, USA, Canada, Australia, Japan

Cover: Foto ©berggeist007 / pixelio.de

More available books at **www.hansebooks.com**

TOBACCO

Curing and Resweating for Quality and Dark Colors.

A PRACTICAL HAND-BOOK

FOR

CIGAR MANUFACTURERS AND LEAF DEALERS

WHO ARE LICENSED TO USE THE PATENTS OF

CHARLES S. PHILIPS,

INVENTOR, AUTHOR, AND PUBLISHER,

188 PEARL STREET, NEW YORK.

CONTENTS.

	PAGE.
Solution for Wetting or Casing Tobacco	5
What makes Tobacco smell bad	5, 7, 11, 12
Why is Ammonia used	5, 7
How to remove bad odors from Tobacco	5, 13
Box for hanging Tobacco into after it is sweated to give it strength, or for Ammonia gas treatment	5, 6
Ammonia proccesses	1, 6
How to sweat Tobacco witnout using Ammonia, and what degrees of heat to use	6
Natural sweating produces its own Ammonia	6, 7
Heats which kill natural sweat or Ammoniacal fermentation	6, 7
Chemicals not necessary to bring out dark colors	7
How much Ammonia should be used	2, 7
Why Tobacco should be packed in tight cases	7, 11
Casing Tub	7
Casing Board	8
How to case Tobacco	8, 9
Casing and heat for tender goods	8
Casing and sweating matted or sticky Tobacco	8, 13
No heat should be used in casing	9
Boxing or bulking for a natural sweat	9, 10
Why blankets should not be used	10
Sweating for quality	10
Repacking into cases for a second natural sweat	11
When Tobacco is in proper condition to be treated for colors	11, 12
Heats that do not take out gum	12
Sweating for rich lively colors	12, 13, 16
Heats at which the apparatus or room should run	12
Havana Seed	12
The cause of mould and how to kill it	13
Why hot Tobacco should not be exposed to the air	13
The cause of dead, red or gray colors	13, 14
Sweating short wrappers	14
Sweating fillers and binders	14
Sweating Havana Tobacco	15
Why clean soft water should be used for casing	15
How to tell hard water from soft	15
Why all Tobaccos should be packed closely together at the butts	15
Heats for sweating Havana	16
Casing Havana for glossy colors	16
New Tobacco	16
The cause, and cure for matted Tobacco	17
C. S. Philips' Patent Portable Sweatting Apparatus	18
How to use it and keep it clean	19
What to do when not in use	20
Natural sweat or fermentation, how killed	20
How to proceed when you are in doubt as to how you should handle your Tobacco	20

INTRODUCTION.

This work is not for the public, but solely for my patrons who have paid for the use of my Patents and Processes, and who are entitled thereby to all my experience and all the aid and assistance I can render them that they may more easily and positively reach satisfactory results. As you have paid for the knowledge which this book contains, it is to your interest that you allow no one not connected with your business to have access to it. It is for your private use only, and only so long as you hold a license from me.

I have made Tobacco my constant study for many years, with a view to improving its quality and sweating it to dark colors; also to perfect a process that would be simple and positive and not require extraordinary skill to successfully use it. During these years I have made the resweating of tobacco to produce DARK COLORS a special business. I have been largely patronised by the leaf and cigar trade in all parts of the land. Many thousands of cases, and large quantities of every kind of seed leaf grown, have passed, and are daily passing through my establishments. I have watched and noted carefully the result of every case. Many costly experiments have been abandoned; prejudices have been slowly overcome. It has been hard, up-hill work, but perseverance will accomplish whatever you undertake; and to-day I have the satisfaction of knowing that I have brought it to such a state of perfection that it is the only successful process in existence, and is in general use from one end of the country to the other, by all classes of tobacco and cigar merchants, from the very largest to the smallest.

In handing you this book I do so with the sole purpose of giving you my experience, and thus aiding and guiding you to successful work.

I would caution you not to attempt too much at the start in the way of using heat. Do not try to use a high heat because you can do so by simply turning on the gas, get familiar with the result of the lower degrees, sweat your first case at 140 degrees; your second 150, and the third at 160, and see which you like best· remember that the lower heats with plenty of natural sweating beforehand gives the most lively and glossy colors.

Have a little patience. Do not expect to learn all relating to the process on the first case you try, but study each case, note the results and make a memorandum, which you can refer to. Be very particular to have all work done in the nicest manner. You cannot put too much labor on your goods; it will all show in the finale result. Read this work until you are perfectly familiar with it. Compare results; you will always find something to learn. The more pride you take in it, the better you will be satisfied.

I have endeavored to be so plain in my instructions that you cannot go astray; and I trust I have succeeded in my object. If in any way I have not, you will please be guided by instructions on the last page, and greatly oblige,

Your obedient servant,

C. S. PHILIPS.

SOLUTION FOR CASING OR WETTING TOBACCO.

Take four (4) pounds of carbonate of ammonia, break it into small pieces, and put it into a barrel which holds 45 gallons. Fill the barrel with *cold* water and stir it until the ammonia is dissolved, then it is ready for use. Do not make up this solution faster than you want it for use, as it loses its strength by standing. It should be made fresh every day or two, and if you are casing only small quantities of tobacco at a time and several days apart, you can make up a smaller quantity of the solution, say one pound of ammonia and eleven gallons of water, and what you have left after you have done casing, you can keep for future use by putting it in a tight barrel or keg, and put the bung in tight; in this way the solution will keep as long as you may wish. The better vessel to keep it in is a carboy, which is a glass vessel holding 10 to 12 gallons and enclosed in a wooden box. They can be had at any drug store at less cost than a ten gallon keg. If you cannot get one, I can send you one. The question will very naturally arise, "Why do you use ammonia?" and without going into any scientific details, and in as plain language as I know how, I will give you my experience in the matter and show you how you may prove to your entire satisfaction by a very few practical experiments, that I am perfectly correct in making the assertion that ammonia is good for our seed leaf tobacco. That in no way is it an injury to it. That, as a matter-of-fact, it is a natural constituent of all well cured tobacco—and that in the proper use of this fact lies the whole secret of producing as dark colored tobacco as you may wish, and of a natural smell and flavor, and without the use of any artificial coloring matter.

There are at least three ways to impregnate or get ammonia into tobacco. My old method was to sweat the tobacco as dark as I wanted it, using only plain water casing, and steam heat from 150 to 200°. This made the tobacco stink, or have the steam or Kentucky smell. I then shook the tobacco out and hung it on pins which were two inches apart. These pins were French wire nails, (1½) one and one-half inches long, driven through a half-inch slat—(2) two inches wide and four foot long—the pins or nails were slanted a little when they were driven in, so that when the slat was put in place, the points of the pins pointed a little upwards, the slat being one-half inch thick and the pins one and a half inches long, left one inch to spare on which the hand of tobacco could be placed. The hands were placed on the pins in such a way that the pin went through the tie of the hand. I then used a large tight box four feet wide, six feet high and ten to twelve feet long; one end was used for a doorway—four inches from the top of the box on each side, I had a one-inch strip two inches wide, the whole length of the box, and another such strip about the middle of the box. These strips are for the four foot slats to rest on. This arrangement allows two tiers of tobacco to be hung the whole length of the box, and it would hold two to three cases of tobacco. In order that the tobacco should not hang too closely together, a blank slat with no pins on it one inch wide and half inch thick, and four foot long, was placed between each slat filled with tobacco. This blank slat kept the tobacco slats one inch apart, and the pins being two inches apart, allowed a free circulation of air all around every hand. After I had hung in the tobacco as described, I placed four shallow iron pans, each having one to three pints of aqua ammonia, in various

parts of the box underneath the tobacco and closed up the door and let it hang there
over night and the next morning I packed it into regular cases. The gas from the
ammonia water would fill the box and be absorbed by the tobacco, and after this tobacco had been packed a few days, the tobacco would have a good smell; the ammonia
having driven all bad smells away. The greatest objection to this process is the expense, as it takes considerable ammonia and necessitates the handling of the tobacco
to hang it up. The process has one great advantage and that is, if fine resweated tobacco is shaken out while it is pretty warm, the leaf is then perfectly free; and if it is
hung up in the manner above described for a few hours, and the door of the box left
open so the tobacco gets a little air, (no ammonia need be used) the tobacco will gain
so very much in strength of leaf that many more wrappers may be cut from it.

The second ammonia process I tried was good in its result, but too expensive.
The tobacco did not require any extra labor in handling it, but it required more ammonia. The process was simply this. I kept the atmosphere of the rooms or sweat
houses, constantly impregnated with ammonia gas by feeding ammonia into the rooms
or apparatus as often as it was necessary; that was determined by the size of the room,
also by the degree of heat used. A pound of ammonia salts or a few pounds of aqua
ammonia would last but a very little while. I also had a large retort heat by steam
and connected to my sweat rooms by iron pipes. In this retort I manufactured my
ammonia gas and fed it into my sweat rooms as they needed it and as fast as they con-
sumed it.

The third and last one I tried is the one I first described, that is. the solution
of ammonia in which I case my tobacco. It has proved cheap and a perfect success.
The ammonia salts which are dissolved in the casing water, is converted into ammonia gas as soon as the tobacco is heated through in the sweat house; and if the case is
tight in which the tobacco is packed, the ammonia gas escapes so slowly that it requires several days to drive it all out, and before it has all gone the colors have become dark enough and the process is ended.

It is not absolutely necessary that ammonia should be added to the water for wetting the tobacco in order to have the tobacco smell of ammonia, or show that ammonia is in the tobacco; for if you wet your tobacco with clean water only and pack it
away a few weeks, and keep the temperature at 75 to 90 degrees, you will find upon
examining the tobacco that it smells strong of ammonia. You will notice also that
all the rank, wild elements of the tobacco have disappeared and the tobacco about
that time shows its best quality. Ammonia also saponifies or cuts up the oils and
fatty substances which were in the tobacco; brightens and enlivens the colors, and
leaves no deposit on the leaf. I have often heard it said that ammonia makes the tobacco gray; that it leaves a whittish or gray powder on the leaf, but such is not a fact.
Ammonia leaves no residue upon evaporation. One thing is certain, and that is, tobaccos that do not show the presence of ammonia, are more rank and wild, and are of
inferior quality to those that do show it. There is such a thing as tobacco getting too
much natural sweat, too much ammonia being developed, and that would cause the
tobacco to become tender and finally rotten. Therefore, the process must be watched.
The thicker and more gummy the leaf, the longer it may sweat; while the thin fine
grades need much less time for sweating.

I have before remarked that ammonia is formed in the tobacco, while it is undergoing a natural sweat at 75 to 90 degrees of heat. I will here remark that, had that
same tobacco been sweated by a higher degree of heat, say 110 to 140, no ammonia
would have been formed in the tobacco—at least none could be noticed in it—for the
reason that the higher degree of heat would drive it out or destroy it as fast as it
could be formed. So you will plainly see what a great mistake it is to case your tobac-

co one day, pack it the next and put it into sweat the next, at heats above 90 degrees, as it takes several days' natural sweating to produce the ammonia and sometimes several weeks.

If tobacco is put into sweat under high heats, 140 and upwards, before it has been cased with the ammonia solution and before it has sweated naturally a few days under heats from 75 to 90 degrees, or before the ammonia has time to fully develop itself, the tobacco will come from the sweat smelling disagreeable; it will have what is called a steam smell or Kentucky smell. This bad smell   was   the cause of all the trouble with the previous process, and was by almost everyone, supposed to be caused by chemicals being used to color the tobacco; but I have shown you that no chemicals are at all necessary to make tobacco as dark as we wish.   Formerly we did not understand that natural or ammonical fermentation or sweat, would drive out and destroy all the wild rank elements of the tobacco which caused the bad smell.   Now that we do understand how to sweat our tobacco sweet and natural in smell and flavor, we have only to follow such simple and natural laws of nature as I have laid down for you to follow, and the closer you follow them, and the more you notice and study the results of each case, the better work you will be able to do and the more pleasure, and satisfaction it will give you in doing it.   The above are my reasons why I use the ammonia in the water used for wetting tobacco by my process; as I can thus do as much in a few days as would naturally require several weeks.   I have shown you that ammonia is developed or made under 75 to 90 degrees of heat, and that all heats over and above 110 dispel, drive it out and evaporate and destroy it.   I have also shown you that, ammonia leaves no deposit upon evaporation; no trace is left behind to injure the tobacco or even show that it had   been used.   So you now know how to make it in the tobacco and how to get rid of or drive it out, and this latter part is quite necessary, for we only want to use enough ammonia in our casing water to last until the tobacco has sweated long enough to bring out the colors as dark as we may wish. And as this requires heats from 140 to 160, it is very easy to see that the ammonia is very rapidly destroyed, and all the pains possible should be taken to prevent the ammonia being driven out faster than is absolutely necessary.   This can be accomplished by packing your tobacco in tight cases.   The tighter they are the less or slower will the ammonia escape and the sweeter will your tobacco sweat.   You will also learn to graduate the strength of your ammonia solution to suit the nature of your tobacco to be operated upon.   Should you at any time have extra rank tobacco, or that which requires an extra degree of heat, say 170 to 190, like Havana seed or ground leafy goods, you would need to use a stronger solution, say 6 pounds of ammonia to the barrel of water.   You will also see that the object is to use just enough ammonia to carry the tobacco through the sweat and have it come from the sweat-room having a natural smell, that is, a slight smell of ammonia.

CASING TUB.

This needs no particular description.   It should be at least as large as an ordinary wash tub and hold at least twenty (20) gallons.   It should be large enough to allow the hands of tobacco to be drawn through the solution in the tub, without the tobacco being forced into such contact with the inner sides of the tub as to break the tobacco. Many tobaccos are so dry that it is necessary to dip them wholly under the water or holding the hands of tobacco by their butts, dip the tobacco tips first, nearly up to the tie.   The casing tub should be of sufficient capacity both in width and depth to allow of the above practice.   The ammonia solution must never be mixed in the casing tub, but always in a seperate vessel or barrel, and dipped from the barrel in which it was mixed into the casing tub.   This precaution will always insure a solution of even strength which is quite important.

CASING BOARD.

The above cut represents a casing board, on which to stand your tobacco after it has been wet by dipping.   It should contain thirty-six (36) square feet; the most convenient sizes are four (4) feet wide and nine (9) feet long, or three (3) feet wide by twelve (12) feet long; this will hold four hundred (400) pounds of tobacco (one case) when it is stood upon its heads, as represented, only very close together.   The side A is eighteen (18) inches high, and straight up and down.   The head board B is the same height, but slants back a little so the tobacco will not fall forward; the side C is six (6) inches high, and is the side on which the caser stands; D is a trough to carry the water or casing into the tub E.   The board rests on two supports F F; the back one should be a few inches the highest to give the board pitch enough to carry any water quickly into the trough D and tub E.

CASING OR MOISTENING TOBACCO.

It is rather a difficult task to say to what extent tobacco should be wet for sweating without seeing the tobacco to be operated upon, and thus being guided by its needs to fully develop it, and by its nature as to what it will stand, and no one will expect me to lay down an infallable rule.

If the tobacco be a good, strong leaf it is a fair rule to bring it back to its marked weight; but if the leaf be thin, fine and large like some of our Connecticut goods, and the tobacco is to be kept for sale, it would hardly be safe to give it so much casing. On such goods it is better to have them run twenty pounds under marked weights than to have them spoil.   If the leaf to be sweated shows a disposition to be tender, and is old goods, a fair casing will not hurt it, providing you watch it, and as soon as it gets fairly warm after it is cased, pack it into your tight case; let it stand forty-eight (48) hours, and then put it into the sweat room, or apparatus.   Run it the first day at a heat of 140 degrees, and the second day at 160, and the third day, and until the colors suit you, run it at 170 degrees of heat, being careful not to run it long enough to exhaust the ammonia and make it smell bad.   Should the goods to be sweated be new goods and show a disposition to be tender, or old goods that are fine leaf and matted so they will not shake out easily, put them cases and all, in their original condition, into the sweat room, or apparatus for forty-eight (48) hours at a heat of 140 degrees; this will warm the goods nicely through, and dry the sap out of the new goods, and soften up the old so that they will be very easily shaken out, and all the leaf will be free and made much stronger than before it was warmed up. Now while the tobacco is hot, shake it out well and sort out any that is not fit for sweating. Lay out each case by itself, in one pile, say eight feet long, butts all one way; this is done so that the pile will not be too high and heavy, so it will cool quickly, and the butts and tips get nicely dried off. Let it lay another forty-eight (48) hours, by that time the tobacco has got cold and you can then case it as you think it needs. No tobacco should be wet or cased while it is hot, the pores of the leaf are open and the tobacco would absorb too much water; neither must warm water or casing be used on tobacco, nor must the tobacco be swung off after dipping or casing; it is only unnecessary labor and does no good, in fact, it does harm by breaking the tobacco. Stand it on the casing board on the heads and all the water will run off that is not needed to sweat it.

If it be gummy and fleshy and soft enough to shake out easily, it needs only enough to make it feel very soft, when it is warm in the pile or case before it goes into the apparatus. In order to reach this result, the tobacco should be dipped into the water, butts first, say about 8 to 10 inches, draw them out of the water at once and let the most of the water run off into the tub again, then raise the butts up straight so the tips of the leaves will hang down, this will allow all the water to run down through the leaves which did not run off the buts while you were holding the tobacco butts down. After you have held the tobacco in this position a few seconds, the water will commence to run off the tip ends of the leaf, then you turn your tobacco butts down again, stand it on the casing board straight up and down. This will allow any excess of water to run off the butts onto the board and into the tub. Keep on this way until your whole case has been cased and stood on the board. Take only as many hands of tobacco into your hands for dipping at one time, as you can nicely reach around with your two hands. If your tobacco should be dry, or a sandy ground leaf, or a very light colored leaf, or an old dead leaf, the dipping should be done heavier, that is, the butts should be put further into the water, and when you draw them out turn the heads up in the air at once, thus allowing more water to run down through the leaf and off the tips. If the tobacco should be very dry, so much so that it cannot be shaken out without breaking the leaf, then it is better to dip it in tips first, up to the ties only, and hold it in the water a few seconds, according to how dry it may be, and then stand it on the board. This standing on the boards allows the water to run off lengthwise of the leaf and perfectly prevents water spots. The tobacco should stand on the boards from two to three hours, or until the tips are so dried off that they do not show an excess of moisture, but not dry enough to be brittle and break. I dip the tobacco butts first, because that part of the leaf next to the butts is the hardest to sweat and needs the most water, the tip part of the leaf being more delicate and easily sweated, requires less moisture and should not go into sweat too wet.

BOXING OR BULKING OF TOBACCO FOR A SHORT NATURAL SWEAT OR FERMENTATION.

After your tobacco has stood on the casing board long enough for the water to get well drained off and the tips show that they are drying off, it should then be taken

from the board and bulked or piled up nicely, so it will go into a natural sweat. I
find it more convenient as well as cheaper and I get a more satisfactory result, by
keeping each case of tobacco by itself. Tobacco certainly sweats better in small quantities than in larger bulks. I therefore take the tobacco from the casing board and
lay it straight and evenly into boxes forty-four (44) inches square and twenty-four (24)
inches deep, inside measurement. If you have seed leaf cases and do not wish to go
to the expense of these boxes, they will answer as well by building them up fifteen or
eighteen inches, this will allow you to put in one such box or case, one whole case of
tobacco laid in loosely, that is, not pressed down any, only lay it in snugly. Keep
the butts close together so they will *not* get too much air around them, and thus be
dried out too much during the few days of natural sweating.

The tobacco should be so laid in that the butts do not touch the wood of the box.
The better way to accomplish this is to make a false head board one inch thick, to
stand in each end of the case while you are laying in the tobacco, and when it is all
laid in, draw out the false head board. This will leave an air space of one inch between the butts and the case; this will prevent the butts moulding while the tobacco
is getting into a natural heat or sweat. The cases should be so handled that the tobacco does not get shook down all to one end. Keep the space always equal on each
end of the case. The tobacco should also be laid into the boxes in such a manner as
to be a little rounded off on the top. Generally, wrappers are long enough so that the
lapping of their tips makes belly enough on the tobacco; but, should the tobacco be
too short for that, enough hands may be laid on the tips to round it off a little. The
hands should all lay the same way in the box. This prevents the tobacco from water
staining on the tips, for if the tips of the leaves lay lower in the box than the butts,
the water naturally runs to the lowest points, and the tips of the leaves require less
moisture than any other part of the leaf. After you have put into a box or case all
you intend to, then cover the tobacco with a wood cover; have it made of such a size
that it will just fit inside of the case. This will keep the top hands from drying out
and if the tobacco should not quite fill the box or case, the cover being small enough
to fit inside will always lay on top of the tobacco, blankets should not be used. They
are expensive and soon get filled with mould, and impart it to the tobacco. Now
these boxes or cases must be set or tiered in a warm room where the temperature can
be kept at about summer heat, that is from 70 to 90 degrees of heat for 5 or 6 days or
until the tobacco gets into a good sweat and becomes hot all through the mass and
under no circumstances must it be disturbed before. This allows the water to become
more evenly distributed through the leaf; do not try to use more heat at this stage of
the process than I have mentioned. It is not necessary that you should try to keep
the atmosphere moist while it is sweating naturally under this low degree of heat, as
the tobacco is wet and packed in tight cases, and what little moisture the tobacco may
lose will come from the butts only and be absorbed by the wood, and would not be
sufficient to interfere with the successful sweating of the tobacco after it is put into
the sweat room or apparatus and subjected to a wet heat. In fact, a little drying of
the butts at this time is rather to be desired, as the butts and ties should come from
the process in not too wet a condition, and if they should get a little dry, they would
become moist enough again as soon as the cases were heated through in the sweat
room or apparatus, by the moisture working from the tobacco toward and around the
butts.

This natural fermentation on sweat determines greatly what the quality of the
tobacco is to be, and the longer this natural sweating is allowed to go on undisturbed,
the better the quality of the tobacco will become, more especially if the tobacco be

Pennsylvania or any other leaf that be heavy, or green, raw and uncured. As soon as the tobacco has got into a nice heat it is ready to repack into cases. The tobacco should then be examined to see if the tips are not too moist. If they should be found to be wet the whole case should be shaken out to see that all the leaves are free and in a proper condition to pack for sweating. It should be laid in one pile for an hour or more so the air has free access to all the tips and butts, and until the tips have dried off sufficiently so the tobacco can be packed without breaking any of the tips. If the proper care has been taken to have the tips well dried off before the tobacco was taken off the casing board, this piling of the tobacco before the final packing will be unnecessary.

The box or case into which you are to pack your tobacco before placing it into the sweat room or apparatus, should be made of wood and sufficiently tight so as to prevent steam or hot water vapor from coming in contact with the tobacco. The idea is to protect the tobacco in any way so as to prevent any of it from becoming over saturated or too wet. By a tight wood box, I do not mean that a box need to be made of matched lumber, as an ordinary seed leaf case in good order will answer every purpose as the wood swells so much as soon as it gets into the wet heat that it is practically tight. Should there be openings in the case not likely to swell enough to come together, cover them over from the inside. The objection to iron, zinc, or other metal boxes is, first they are expensive and the gases from the tobacco destroy them very rapidly; secondly, they cannot be handled while they are hot and wood boxes can be. The process or the result would be the same in a metal box. I have seen it stated that iron or metal boxes in contact with the tobacco tainted it and gave the tobacco a bad smell. Such is not a fact. The tobacco was made to smell bad from over heating it or heating it too long—140 degrees of heat will do so in time—210 degrees will do so in a few minutes.

In packing into cases take plenty of care to lay the hands in very nicely; two to four hands at a time, according to the size of the hands, lay them in very straight and let the butt ends come snug up against the head boards of the case you are packing into. Pack it snugly together; put as many hands as possible in each layer every time you put in a layer across the case. The idea is to pack the tobacco so closely together that the vapor or steam has but very little chance to get around the tobacco. In fact, if the cases are so tight that no steam or vapor whatever can get into them at the tobacco, so much the better; all we want is, such a moist atmosphere around the cases that the tobacco does not dry out. The tobacco has all the necessary moisture from the casing; we neither wish to add to it or take any way, consequently the tighter and more perfect are the cases into which the tobacco is packed to go into the apparatus, the better will be the result. After your tobacco has been packed from the boxes into cases that you mean to put into the sweat room or apparatus, you must again let it lay in cases and in a warm place at 70 to 90 degrees of heat as before described, until each case gets heated through again, which will require from two (2) to four (4) days. Now is the time to decide when the tobacco will be fit to go into the sweat room or apparatus.

If the tobacco is well cured and *smells good* and strong of ammonia, it is ready to be finished off in the apparatus; but so long as the tobacco shows a green, raw and uncured condition and does not smell good, and swells it *must not* go into the apparatus, but must be left to a natural sweat until the ammonia has driven out all rankness and a good flavor is established. If you will give tobacco a little extra care at this stage of the process, follow the rules I have given you and have a little patience. You will be well repaid for all the trouble you may have had by being successful every time. Tobacco cures better under a low heat, that is, more gum is thrown out or decomposed

TOBACCO CURING AND RESWEATING.

by heats of 90 or 100 degrees than by the higher degrees; so if you have a leaf that swells, you must get the swelling out by sweating out the gum by fermentation or natural sweat by using a low heat as above specified. Heats from 120 up to 180 do not throw off any gum, at least not enough to make a swelling leaf burn good during the short time which tobacco can be exposed to such high heats without spoiling the leaf or its strength. Such heats act more to color the leaf a dark color, than to rid the leaf of any of its gum. So you will see why a leaf that has not much gum or none to spare, needs but little natural sweat and a quicker sweat at the higher heats which bring out the color and which run from 140 to 180. At a heat of 200 to 210, the properties of the leaf are decomposed or changed into an oil called Empyreumatic oil, which gives the stink to tobacco, called by the trade a Kentucky smell. From the above explanation of the action of the different degrees of heat on tobacco, you will very readily see why you should keep enough tobacco cased ahead so as to give it all the time necessary to get it well cured before you attempt to finish it off for colors in the apparatus. No tobacco should go into the apparatus that has not been cased and in a natural sweat or heat from 6 to 10 days. The better cured your goods are the less time it will take in the apparatus and less expense, and the better will be the result. If you want good rich, lively colors, you must pay particular attention to the above facts. It is just as easy to keep a few days stock cased ahead, as it is for one day. Some tobaccos can be made ready in six days. For instance, fine thin leaf old Connecticut that has no gum to spare, a little experience will enable you to tell exactly all about it. The most particular care should be used in sweating long fine Pennsylvania wrappers. They will not color nicely unless they have been well cased and well sweated before they go into the apparatus. To satisfy yourself that I am right, you experiment a little this way. After your tobacco has been cased and put into piles or boxes and allowed to remain so for about six days, and then packed into cases, you take one case that has been packed three days and put it into the apparatus to finish it off for dark colors, and make a memorandum just how many hours it is in the apparatus before it gets dark enough to suit you. Now take another case that has been packed six days, and then another that has been packed nine days, and you will know how much shorter time it requires the older cases to sweat and how much nicer the colors are than the first case of the lot you sweated. No positive rule can be laid down for sweating tobacco that would apply to all cases. Each one must be handled just according to its own peculiarities. The older the tobacco is the shorter and quicker the process should be; and the newer the tobacco, the slower and longer it should be and the less heat you should use. Now that the tobacco has had natural sweat enough, it is ready to go into the room or apparatus to be finished off for dark colors. If the goods are what we would call new goods, run your heat at 140 for three or four days, then examine them to see if they are dark enough. If they are not dark enough and smell good, then run them one more day at 150; but if they do not smell good, do not increase the heat. If the goods should be old ones, run the first day at 150 and two days at 160. Examine them and if not dark enough, finish off at 170. When running over 160 degrees of heat, the goods should be examined night and morning. After they have been in the process over the third day, this is so that they shall not be exposed to the high heat over twelve hours at a time without being examined to see if they are done.

Havana seed should be cased and naturally sweated the same as other goods, but needs more heat to bring out the colors. While in the process or sweat room use heat as follows: First two days, 140; third day, 160; fourth day 180 and finish off at 180 or 190 degrees of heat, being careful not to prolong the process more than is actually necessary, as such heats are apt to produce a deadish black color; but if examined often while using such high heats, say every six hours, the darkest colors may be reached and yet have a lively appearance. Great care must also be taken that the tobacco is moist enough to allow dark colors to be produced or brought out. If the tobacco should be too dry a greyish color will be brought out. It is very easy to tell whether or not the tobacco is moist enough, by drawing a few hands from the mass under treatment and shaking them out a little in the open air. If the tobacco almost instantly assumes a harsh and brittle nature, it is too dry to get the colors and there would be no use in further continuing the process; but shake the tobacco all out, let it dry off and case it over again. You will very easy reach the colors the second time. Should you find it necessary to prolong the process to such an extent as to drive out all the ammonia and produce a disagreeable or objectionable smell on the leaf, you can remove the smell by hanging the tobacco and treating with ammonia as before described, or simply pack it up and let it stand a few days and the bad odor will disappear; or, if the leaf is strong enough to stand it, you can hang the tobacco until it is dried off pretty well and then case it over again, and as soon as natural sweat or fermentation sets in all objectionable odors will entirely disappear, which again proves to you how beneficial natural sweating or fermentation acts upon tobacco.

Upon examining any case and finding it dark enough to suit you, you of course, take it from the process and set it one side to cool off somewhat, if not dark enough continue the process. But when you put the cases back into the room or apparatus, reverse it, so that the side of the case which was up so far during the process, should now be down, as the side which was up will probably show that it has colored a little the best. This difference will not be quite so apparent where my single case apparatus is used, as the space to be heated is so small, there is hardly if any difference of temperature in the apparatus. After any case has been standing out of the process an hour or two, it must be nicely shaken out, then you can at the same time repack it into its original case, if you so wish, as fast as you shake it out. If the tips or butts are too moist, shake it out and let it lay in a pile a little while before repacking it. If you pack it again into its original case, use a false head board as described in boxing the tobacco, also take out one or two head boards from each end of the case and leave them out. This will leave an air space around the butts and allow them to dry off. Tier these cases in a middle tier, so plenty of air can get at the butts, and keep them from moulding. If your tobacco should mould, it is because the air around the tobacco is not dry enough. Should you have any mouldy tobacco and moist, and strong enough to stand it, you put it into the sweat room and heat it 48 hours at 170. This will generally kill mould, unless the tobacco be dry then there would be no use in trying it. If you do put the case in to kill the mould, it is to be shaken out and repacked afterwards the same as the other goods that go into the process.

The reason why the tobacco should be allowed to stand some little time to cool off before it is shaken out, is: That if at once, in its hot condition, it was to be exposed to the air, it would loose much of its moisture and the leaf would be inclined to thicken up by the pores of the leaf suddenly contracting. It should be shaken out nicely before it gets cold, as in its warm state, every leaf is freely and easily opened by the shaking, whereas, if left to get cold, much of the leaf would stick together and soon be like plug tobacco. In winter weather no hot tobacco should stand over night without being shaken out. Should a case get cold and sticky, warm it up again at 140 degrees of heat.

As a rule tobacco that has been in the apparatus under a moist heat four or five days at an average temperature of 145 degrees should come out a RICH, GLOSSY, DARK TURAL SMELL. If the leaf should then be of a RED, GREY or DEAD COLOR, or SMELL BADLY it is because you did not WET the tobacco enough or did not give it long enough time in the NATURAL SWEAT. If the tobacco becomes quickly dry and husky when you take it hot from the case and shake it in the air, then you did not case it enough, but if it remains moist and stretchy, then any bad or unsatisfactory result is caused by you not allowing the tobacco to sweat naturally under a low heat, long enough before you put it into the apparatus.

SWEATING SHORT WRAPPERS.

The shorter the leaf or the less length there be to it, the less the heat required to produce dark colors. 140 degrees is a plenty for B and C stock, or anything less than a single A, and 72 hours is the average time where the heat is kept uniform night and day; and if you are using a sweat room where you tier your cases in the room two high, you should *always* place on the bottom or floor of the room all the cases which contain the *shortest leaf*, and all the cases which contain the *longest leaf*, should be tiered two high, as the upper part of the sweat room will naturally be a little the warmest; as heat will rise to the highest point of the room.

FILLERS AND BINDERS.

As a rule, fillers, or binders, or sandy ground leaf goods, should not be re-sweated, except for immediate use as soon as they come from the process; they cannot then be kept with any degree of safety; they very easily run into mould and rot. But if you do wish to sweat such goods, and hold them for sale, and do not know just when they will be worked up, the following rules or precaution will prove useful: Do not attempt it on new tobacco, that is, goods that are not several months old in their cases; they should be what the trade would call old goods. Do not case them too heavy; as soon as they get warm rehandle them and pack them and let them stand 48 hours in their cases, and then put them into process at 140 degrees. Sweat them the same as you would short wrappers; when they come from the sweat, be sure you do not pack them too wet; they should be be just nicely soft without being wet. Should they feel pretty soft hang them up a little while, as I before described, in the ammonia boxes. The dryer you pack them, of course the more surely will they keep. If you pack them for sale, the cases should be cut down lengthways, so as to make the case narrow enough to allow the tobacco to be packed in with the butts all against the sides of the case instead of the ends, and the tips just nicely lap on each other; use a false board, as before described, so as to leave au air space between the butts and the wood of the case. Also, leave out a side board on each side so as to allow plenty of air to get around the butts to dry them off. This mode of packing allows every butt to be exposed to the air, and not to be packed in cross packed, the way most short goods are packed into seed leaf cases, as the hands in the cross packing cannot dry out fast enough, they first mould and then rot, and being in the middle of the case, and lying across the tips and finest part of the leaf of the balance of the case, the whole case very soon has caught the disease and is worthless. Do not cut the case so narrow as to make the tobacco belly up too much in packing it.

HAVANA TOBACCO.

The resweating of Havana tobacco by the use of high heats. is not to be recommended, it does not agree with the leaf, neither does the leaf require it. The leaf is so short that it requires but little heat to develop it sufficiently as the taste for colors in Havana goods runs more in the rich dark brown shades, and these colors may be brought out nicely by natural sweat or fermentation, after being properly cased. In the first place, the water used for casing should be soft, for the reason that the coloring matter of the leaf is more soluble in soft water than in hard, and if the water in your vicinity is hard, you should use rain water for casing, or have your hard water treated chemically to render it soft. This can be done at a trifling expense. Rain and snow waters are the purest kinds of natural water. A good water may be known by its being fit for cooking purposes, and will not curdle soap. If your water is hard it will curdle soap; but if it be soft it will make a lather of the soap. Some think stagnant water the best for casing tobacco; I do not. Water becomes stagnant on account of its impurities, and they act as ferments. All ferments have a certain life to live like everything else and then die, and I much prefer to use the water fresh, and let all the changes of fermentation take place while on the tobacco; therefore, in casing Havana tobacco, I use fresh or newly drawn soft water. If the tobacco is very dry, dip the whole carat under the water for a second or two and then stand it on the casing board, tips up, to drain off and soften up. As soon as the tobacco gets soft enough so the strings can be taken off and the carat loosened out, the hands taken apart without breaking the leaf, it is then ready to be cased. Now take three or four hands of tobacco in your hands at a time; take hold of them by their heads or butts as they are most generally called. and dip the tobacco into your casing tub, *tips first*, and nearly up to say within one inch of the tie. Now if the tobacco be pretty dry hold the hands in the water a little, long enough to count one, two, three. If the tobacco should be soft, dip in the hands and draw them out at once, and in either case, when you draw the hands out, *swing off* all the surplus water and stand the tobacco on the casing board the same as I have described for seed, enough water will run down on to the butts from the leaf to moisten them all they need. When the tips, in an hour or so. begins to show dryness, then take it from the board, one hand at a time, and pack it straight and snugly into a box made expressly for it. So no butts get covered up in packing it, the same as I described for sweating fillers. Now that you have it all packed into a box. fit a wood cover down on to the tobacco, press it down with the hands only, so there will be no belly on the tobacco and fasten down the cover. Now set the box in a warm place as described for the natural sweating of seed leaf. After it has stood for four to six days, examine it to see how it is getting along, and if the tips should be too wet and sweating too fast, pile it all out for an hour or so to let the tips dry off somewhat, and repack it the same as before. The more pains you take to pack it straight and snug or closely together at the butts, the better will the heel of the leaf sweat and color. The same rule will apply with the same force in packing seed leaf. Havana tobacco is very apt to mould after it is cased, especially old goods that have but little gum and can only be sweated with safety for immediate use. If you discover mould spots on the leaf (and it should be examined for them every few days until it has sweated enough to suit you). You must then put it into your sweat room or apparattus at once and heat it through at 140 degrees, which will require twenty-four (24) hours, then shake it out and use it up, the heat will kill the mould and prevent its farther progress or development. Should you wish to sweat it quicker than the natural process, you can put it into your sweat room or apparatus, at 120 to 130 degrees of heat. This applies to old goods only, after they have sweated naturally a few days. Should the goods be new, they must sweat naturally until they are well cured before they can be put into artificial heat, the same as new seed leaf. If your Havana be old and of a dead nature and a dull color, and you wish to give it a more lively and glossy appearance, you put into ten gallons of pure or clean water two ounces of pure glycerine, and use this solution for casing. Should your tobacco be very heavy in leaf and so gummy it swells, and it needs a good strong sweat, you can use the following solution to case it in: Water, ten gallons; malasses, one quart. Any kind of molasses or syrup or the same quantity of any kind of sugar, will answer the same purpose. The tobacco which this solution is used on must sweat naturally 10 or 12 days, then the sweet substance or saccharine matter, you will find, will have wholly disappeared, and acetic acid is then in the tobacco in place of the sugar or molasses, and the tobacco will have a pleasant sour smell; two quarts of cider or cider vinegar or sour grape wine of any kind added to the above solution, will give the tobacco a somewhat better smell, but will not improve the tobacco in any other way.

NEW TOBACCO.

I have tried a great many different ways to cure new tobacco, that is, tobacco that has only been dried on the poles after being gathered, and then packed into cases, but I have so far perfected no process that does away with natural fermentation, yet I hope to some time in the future. I do not mean to say I have made no advancement in curing new tobacco, for we can now manufacture the leaf a year or more sooner than was formerly done. This year 4 acres of Massachusetts Havana seed was cut the fore part of Sept. 1879; hung in the shed, dried, stripped, packed and shipped to me. It was then put into my process, sweated for dark colors and went into the manufacturers hands and worked up in March 1880. The best plan of procedure which I can now give you, is as follows: After the tobacco has been gathered from the field and cured in the sheds, stripped and packed into cases a few days, take a head board from each end of the case and place the cases in a room where an even temperature of 70 to 80 degrees may be kept up, this heat may be dry, so as to suck the moisture from the butts and the large middle vein of the leaf. The heat must only be sufficient to keep the tobacco in a natural sweat and dry enough to keep the butts drying out, until the big vein for 6 inches in the case gets as dry and brittle as a pipe stem. This will take some time.

Much will depend upon how heavy the leaf is, and the quantity of green sap in the stem. You will have to decide that after it has been under treatment two or three weeks. You can also tell when you are using too much heat by the smell the tobacco has. Green tobacco is very easily decomposed by even low heat; it has an unnatural smell; a stink, a Kentucky smell, which you ought to easily distinguish from the smell of natural sweat. So you will see the importance of not trying to force the sweat too much on new tobacco. If you do, you only delay the process; for let the heat get too high and a bad smell once established, you can only get it out again by a good ammonia casing. Even then it takes a natural sweating of many days to wholly get rid of it. When your tobacco has sweated this way long enough, which you can tell by trying the burn to see if it swells, and when it is dry enough, as you can tell by the butts and stems, then put the cases into the sweat room or apparatus for forty-eight (48) hours, at a temperature of 130 to 140 degrees.   The atmosphere being wet either by steam or water vapor, the tobacco will thus become soft enough to allow it to be taken from the cases and shaken out without breaking any of the leaves.   At the expiration of the 48 hours, shake it out nicely, sort out all the poor or "off" hands and pile each case by itself in long shallow piles, seven or eight feet long, with the butts all one way and let the tobacco lay thus with the tips and butts exposed another 48 hours.   It will then have become cold and can then be cased with the ammonia solution according to its needs, and must then be sweated naturally until all rankness has disappeared, and is to be further governed by the rules before laid down for casing, boxing, packing, sweating, etc.   If your goods are a fine, thin texture, you should not, originally, pack them so heavy as to make them cure dark and matted in the center of the case.   It should not be pressed in with a press.

If the finest Connecticut tobacco was only packed with less moisture in it, so it would not sweat so much during the warm weather, it would then reach the manufacturers in such a condition that it could be sweated as dark and safely as any other crop.

It is not unusual to find 25 to 50 pounds of tobacco in a case of fine wrappers wholly unfit for wrapping purposes, just because they were packed too heavy.   As a rule, wrapping cases do not yield much more than one-half to two-thirds what they ought to, and all from the mistaken policy packers pursue in packing their goods, both too wet and too heavy; so they will sweat hard during the warm months.   Of course some of the tobacco in a case sweats dark, but only the middle hands, and they get so tender they are of no use.   Tobacco will not stand such long periods of moist sweating without spoiling the leaf.   Curing the leaf and sweating for dark colors are two independent processes and cannot both be done at the same time and bring out a strong leaf.

The packer should not stop to think what color his leaf will be when it is cured, only to cure it in such a manner that the manufacturer gets a strong leaf that he may manipulate it to suit his trade without injury to the leaf.   He should pack it in such a state of moisture that when warm weather comes it will go into a sensible fermentation or sweat and thus bring up the quality; but sweating for colors should only be done just previous to its being manufactured.   Curing must be a drying process, and coloring a wet process, and so long as the leaf contains moisture it is constantly undergoing a slow decomposition until the leaf dries out or rots.

C. S. PHILIPS' PATENT PORTABLE SWEATING APPARATUS.

CUT No. 1.   CUT No. 2.

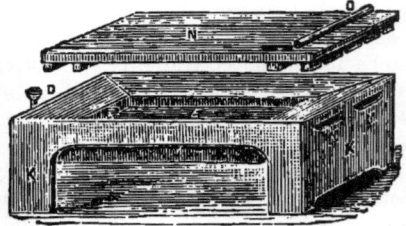

DATE OF PATENTS.

September 26, 1876.
March 12, 1878.
December 9, 1879.
June 15, 1880.
November 9, 1880.

Cut No. one represents the apparatus complete, as it looks when in use. It is 4 feet long, 3 feet wide and is 5 feet high, being just large enough for one original case 400 pounds of tobacco, case and all. The roof has sufficient pitch to carry the water of Condensation to the back end of the house and into the pan E again where it first came from.

A, is a water tank which sits on top the house, or it may be placed on a bench on the floor so the faucet B will be over the water pipe D.

B, is a faucet.

C, is a pipe to carry the water from B to D.

D, is a pipe to carry the water from C to pan E, which is in the bottom of the apparatus.

E, is a metal pan in which water is heated and is connected with the water tank A by pipe D and C, and must always have enough water in it so it may be seen from the outside by looking in the tin funnel at D.

F, is two gas burners underneath pan E, or an oil stove may be used.

G, is the door of the apparatus.

H, is a thermometer on the door.

I, are the six fasteners to keep the door in place.

J, is the top part of the sweat house.

K, K, is the base, 18 inches high on which the top part J rests, and in which is the water pan E. See Cut No. 2.

L, L, are the handles on the door G, by which it is lifted out or placed in position.

M, M, are air holes from the burners, and around the end of pan E.

Cut No. 2 represents the base of the apparatus K, K; with the house J, taken off, also the interior construction. E, is the pan, D, is the water supply pipe for the pan.

N, is the floor of the house, which is just over the pan, but in the cut it is raised up high enough to show the hopper shaped bottom around the pan E. This bottom is made that shape for the purpose of allowing all water of Condensation to run back to the pan where it first came from, thus nothing is wasted.

O, is an iron roller on the floor N, to roll the case in and out on, and should be left in under the case.

Every apparatus is put together and tested before it is shipped. In setting them up look at Cut No. 1, which shows the apparatus complete; put the tin funnel in the water pipe shown at letter D. You will find screws for putting the apparatus together and screw holes to match. No nails are use in its construction. Any time the handles do not close the door tightly, turn up the nuts until they do. Between the pan and the wood work is a layer of fire proof material (asbestos). *Each burner* will consume seven and one-half ($7\frac{1}{2}$) feet of gas per hour when running under high heats, and each large apparatus should have a *one-fourth* ($\frac{1}{4}$) *inch service cock*, and where more than one of these apparatuses are in use, there should be $\frac{1}{2}$ or $\frac{3}{4}$ inch pipe from the meter to the $\frac{1}{4}$ inch service cock.

The whole apparatus is built of wood, with the exception of the pans, pipes, handles, &c., consequently it cannot radiate any heat. It is tight so no vapor can escape, and you could use it in your office and hardly know it, only you could see it. Heat cannot come through wood in sufficient quantity to heat the air outside the sweat house.

It being portable in every respect, it can be placed anywhere you want to do your work, whether it be in a cold cellar or a hot loft. It needs looking after only once in 24 hours, and runs night and day alike. The heat will not vary after it is once established. To opperate the apparatus, you first fill the water tank A. with water, then open the faucet B, the water will run through pipe C into D, and into pan E. When enough water has run in the pan it will show itself in the tin funnel which is in the opening of pipe D, then shut the faucet so it only drops into pipe C, or funnel D, about 100 or 125 drops a minute, so as to keep the water always at the same level; now take out the door G, and head your case up squarely in front of the apparatus, and about two feet from it, and so that it goes in on the flat, tip the case over and into the apparatus, so that the case will rest on the roller, now lift up the other end of the case and at the same time push it into the apparatus, then put in the door. Now turn on the gas and light the burners. Be sure that the burners are *only* lit at the top. *If they should take fire in the round holes at the bottom of the burner turn off the gas and light them over.* The top of the burners should be about one inch from the pan. If the thermometer shows too much heat turn off the gas a little.

It being necessary to examine the tobacco while in the apparatus, to find out if it is dark enough, you simply take out the door G, draw the case out about one foot, raise a board of the case, and draw a hand or two. If not done, close the case, push the case back and put in the door again. Do not interfere with the heat unless the goods are done and you wish to stop the process.

There are two sizes of these Sweating Apparatuses; the larger one for one original case of 400 pounds; the smaller one for 100 pounds at one time. A less quantity may may be sweated in either machine.

As a matter of considerable economy you should keep the water pan clean. Wash out the inside occasionally and brush the soot off from the underside. Take off the burners and clean them thoroughly with a good stiff brush. Do not let them get clogged up by the soot which falls from the pan; if the burners smoke and make soot it is because they are burning at the round holes, or you have too much gas turned on. The cleaner you keep all these parts the less fuel or gas will be required to do your work. Place a piece of sheet iron or zinc on the floor under the burners, have it *two* feet wide and *three* feet long. This is to catch any soot that may fall from the pan If you use an oil stove let the top of the stove just touch the bot-

tom of the pan. Use 130 or 150 fire test oil. You must not use a poor quality. The
best is the cheapest and the safest. Keep your stove wicks trimmed straight across
and do not turn them up high enough to cause them to smoke. Fill your stoves and
trim the wicks every afternoon immediately after dinner. Then you have day-light
for your work and they are in good order to run all night. The safest plan where
oil is used, is to set the apparatus in a cellar which has a cement floor; but if you
must use it on a wood floor then you should lay a course of brick and mortar on the
floor about one foot larger all around than the apparatus is, on which it may set. This
will prevent any oil being spilled upon the floor. Be careful to use clean water in
your water tank so the faucet will not get stopped up. Open the faucet wide occasionally for a few seconds, and thus be sure there is no dirt collected in it. This
should be done just before leaving it for the night. In putting a case into the process, see that it is not done in a rough and careless manner, especially in pushing it
back into the apparatus after it is once on the roller, as it is rolled back very easily;
and if you let the case strike to hard against the back end you may injure the apparatus. Any time the apparatus should not be in use be sure the *water pan* and *water
tank* are both full of water and the door of the apparatus closed. This will keep all
the parts moist and prevent shrinkage.

NATURAL SWEAT KILLED.

Natural fermentation or sweat is killed or arrested for long periods of time by
using certain degrees of heat, or any degree of heat from 140 degrees upwards. This
will prove useful to you if you have fine thin goods which needs some sweating, and
yet do not need a heavy sweating, and where you want to hold the goods for sale,
and do not wish them to continue sweating naturally. I will illustrate what I mean:—
Allow that you have two cases of tobacco as near alike as they can be, and you case them
as near alike as you know how, and pack them up. Now in this condition these two
cases would sweat all summer naturally, and before the winter cold stopped them they
would probably be spoiled; but you take one of the cases after it had been cased and
packed a few days and heat it through, say 48 hours at 160 degrees temperature, then
shake it out and repack it, and place it along side of the other case, and I can almost
guarantee it will not go into sweat again no matter how hot the summer may be. You
now shake out the case you did not heat up and repack it, and in 48 hours you will
find it hot and sweating as much as before you repacked it. This applies to old goods.
Should the goods be new and 160 degrees be likely to make them smell bad, use as
much heat over 140 degrees as you safely can and not bring out any bad odor; try 140
then advance a few degrees at a time until you have gone as far as you can, and keep
the goods sweet and natural in flavor, and not go over 160 degrees of heat.

Finally, any time you feel in doubt or undecided as to the best course to pursue
in order to get the best and most satisfactory results, I would be pleased to have you
send me a hand of tobacco drawn from the centre of any such case, I can then return
it to you with such instructions as you may need regarding it.

Neither dry off nor moisten any sample you intend for me and pack it up in such
a way that it will not dry out while on its way to me. I will then get it in its original condition. Also let me know how old, and what kind of tobacco it is, the marked
weight and tare, and reweight; that I may know how much it has lost in weight.
This will cost you but very little, and may save you considerable trouble. I am

Your obedient servant,

CHAS. S. PHILIPS,

188 Pearl Street, New York, N. Y.

www.ingramcontent.com/pod-product-compliance
Lightning Source LLC
Chambersburg PA
CBHW031158090426
42738CB00008B/1387